Rock Identification Made Easy

By Patrick Nurre

Rock Identification Made Easy

By Patrick Nurre

Rock Identification Made Easy
Published by Northwest Treasures
Bothell, Washington
425-488-6848
NorthwestRockAndFossil.com
northwestexpedition@msn.com

Rock Identification Made Easy
Table of Contents

Rock Identification Made Easy
Introduction

Kids will find all kinds of rocks and they will find them in the most unlikely places! This is what they are exposed to day after day. But studying rocks is not the end of the story. In order to understand rocks, you have to study minerals along the way. So in this study we will not only study the rocks but the minerals that make them up. Minerals like quartz, mica and feldspar are common and will probably be what your kids discover in your driveway or along the sidewalks. Most minerals are rare and are often hard to identify in the field because they most often occur in their alternate cryptocrystalline (hidden) form, not in the beautiful crystal shapes that are found in museums.

Rock identification can be very fun and valuable for learning, but only if kids (and adults) figure out how to identify them. This is where most guides fail, in my opinion. They use illustrations that are inaccessible to kids; only found in museums. They use language that is difficult to understand; only trained geologists can decipher their meaning. This greatly frustrates kids and they soon become disinterested. My goal is to help all learn some simple, basic principles in identifying the wonderful rocks all around us.

Teacher Notes

The study of rock types is the foundation of geology. It is fascinating how rocks demonstrate order and complexity. For example, why are certain crystallization processes in the rocks nearly always consistent? The fact that we can categorize types of rocks demonstrates that we have an orderly God who upholds "all things by the word of His power," (Hebrews 1:3).

As we search for answers to the history of our earth there are really only two explanations for the rocks we see. The first explanation is a 'Johnny-come-lately' idea called Uniformitarianism, popularized by Sir Charles Lyell in the early 1800s. His work was the primary influence in turning Charles Darwin away from a Biblical explanation of earth history. The goal of Lyell was to eradicate from geology any mention of or influence from the Bible. This doesn't sound like the open-minded scientists we are taught to believe in! This particular view itself was a product of The Enlightenment, a philosophical period of history in the late 1700s and 1800s characterized by an unfounded turning away from a Biblical God and His values. Modern, so-called scientific explanations of the history of the earth have been influenced by this philosophy and really are based on the ***assumption*** of Uniformitarianism - an unfounded bias against the Biblical view of earth history. Simply put, uniformitarianism says that what we observe in the present is the key to explaining what happened in the past.

The second view of earth history centers on the great flood mentioned in Genesis. For several hundred years this view formed the convictions of hundreds of scientists. So, when we study the rock types, we are really asking ourselves from where these rocks came. Which is the better explanation? Which explanation best fits the rocks? Which explanation best fits with what we believe about the Scriptures? The great scientists such as Newton, Faraday, Kelvin and many others thought the Biblical view of earth history was a satisfactory and fulfilling explanation.

Although this study can be done independently, if your student is younger (K-3rd) it is necessary that you do this study alongside your student. Please read this next section to familiarize yourself with the concepts that will be presented in the text.

Some organizing principles in rock identification

There are just a few principles to learn in order to make rock identification easy and fun.

1. Learning how to identify whether a rock is **coarse-grained or fine-grained.** In a coarse-grained rock, the grains of minerals can be seen with the naked eye. In a fine-grained rock, the grains of minerals cannot be seen with the naked eye. Only a general color can be seen. Here are some examples.

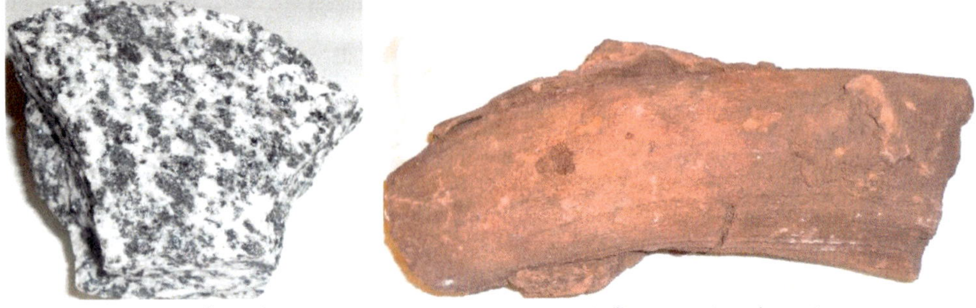

A coarse-grained rock A fine-grained rock

2. Learning just 12 rock-forming minerals. Although there are over 4,500 recognized minerals out there, most of the rocks on our earth are made predominantly of just 12 - 6 light colored and 6 dark colored minerals. It is these minerals that give rocks their colors.

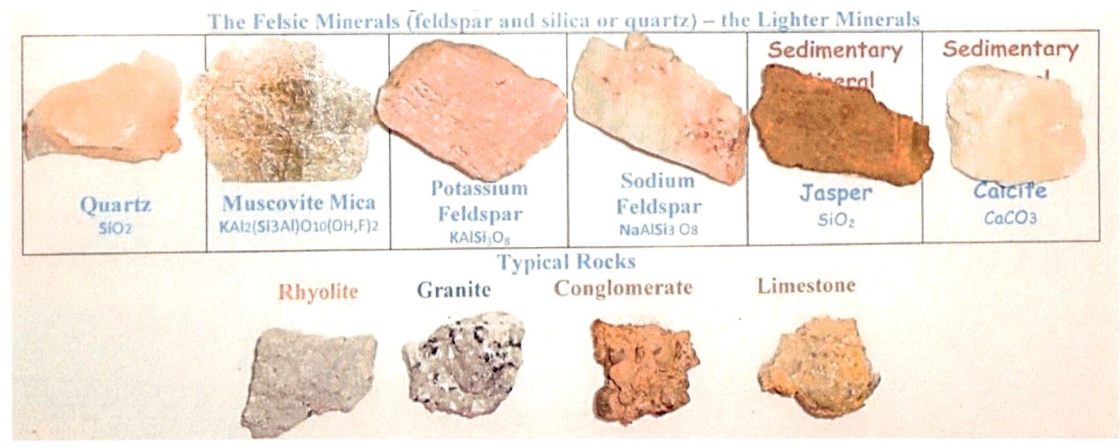

The 6 lighter colored rock-forming minerals; they form lighter colored rocks like the four examples at the bottom of the chart

The 6 darker colored rock-forming minerals: they form darker colored rocks like the two examples at the bottom of the chart

Classification of the rock types

Secular geologists classify the rocks according to how they *think* they formed. They divide rocks into three rock types this way:

1. **Igneous rocks** - those rocks they think formed from fire or heat over hundreds of millions of years. The word igneous comes from the Latin word for fire. We get our word 'ignite' from this word.
2. **Metamorphic rocks** - those rocks they think formed through heat and pressure deep underground over millions of years and then were slowly pushed up to the surface of the earth. Our Biblical word in Romans 12:1-3 is the word metamorphose - "...do not be conformed to this world, but be transformed (metamorphosed) by the renewing of your mind...."
3. **Sedimentary rocks** - those rocks they think formed through gradual deposition and erosion of sediments over millions of years.

A young child interested in learning about rocks will most likely have learned this classification system even in Christian school. But let's take a second look at this.

Has anyone ever seen granites or gneiss form or in the process of forming? The answer is a resounding, "No." Then why do geologists insist that they know how they were formed? Very simply; since the Biblical view of the creation and flood have been rejected by modern science, the only alternative is an ancient earth that has been unfolding for 4.6 billion years. It's that simple.

So, I would recommend that your children learn a different system of classification involving four rock types. It will look something like this:

1. Plutonic rocks, named after Pluto, the god of the underworld. Secular geologists also refer to them in this way. Plutonic rocks are also called 'coarse-grained' rocks because in these rocks one can see the mineral crystals without aid of magnification. These would include the following and their varieties:

- **Gabbro** – a dark colored rock
- **Diorite** – an intermediate colored rock
- **Granodiorite** – an intermediate colored rock
- **Granite** – a light colored rock
- **Pegmatite** – a type of granite with large mineral crystals

2. Volcanic rocks, named after Vulcan, the god of fire. Incidentally, with the exception of a few sedimentary rocks, this is the only rock we have seen forming. Of course they come from volcanoes. Volcanic rocks are also called 'fine-grained' rocks because you cannot see the mineral crystals with the naked eye. Volcanic rocks would include the following and their varieties:

- **Basalt** – a dark colored rock
- **Andesite** – an intermediate colored rock
- **Dacite** – an intermediate colored rock
- **Rhyolite** – a light colored rock
- **The pyroclastic rocks** – volcanic materials that have been ejected from volcanoes

3. Metamorphic rocks, as the name implies, appear to be changed rocks in some way. These rocks appear to have once been plutonic, volcanic, sedimentary or even metamorphic rocks. They look like they have gone through some significant changes involving heat and pressure. But because we have not seen them form and have not seen them in the process of forming, we can only guess as to their origin. Metamorphic rocks would include the following and their varieties:

- **Gneiss** – a coarse-grained rock appearing as **banded alternating light and dark layers**. Some think that gneiss might have come from granite or some other plutonic rock.
- **Schist** – is a **layered**, coarse-grained rock appearing as though someone sprinkled glitter on glue. Some speculate that schist is metamorphosed

shale or mud with the addition of mica flakes. Of course, we don't know for sure.

- **Slate** - is a **layered**, fine-grained rock that looks like shale, only much harder. Some think it is probably metamorphosed shale.
- **Phyllite** - is a **layered,** fine-grained rock that looks like slate but with a very shiny surface. Some think it is metamorphosed slate.
- **Marble** - a **crystalline** appearing rock. Some think it was originally limestone, as it has an abundance of calcium carbonate.
- **Quartzite** - a **crystalline** appearing rock. Some think it was originally sandstone, as it is made up of quartz.
- **Serpentinite** - a **crystalline** rock that has shades of green, which is how it got its name. It can appear chunky or layered.

4. Sedimentary rocks, as the name implies are made of fine-grained or coarse-grained sediments, cemented together with quartz, calcite or even iron oxide. Sedimentary rocks would include the following and their varieties:

- **Shale** - a **fine-grained** sedimentary rock made of very fine broken particles of clay, cemented together. It is also called a clastic ('broken') sedimentary rock.
- **Siltstone** - a **fine-grained** sedimentary rock made of silt-sized broken particles of sedimentary rock, cemented together. It is also called a clastic ('broken') sedimentary rock.
- **Sandstone** - a **coarse-grained** sedimentary rock made of quartz crystals, usually rounded, cemented together by silica or calcite. It is also called a clastic ('broken') sedimentary rock.
- **Travertine** - a **fine to coarse-grained** chemical sedimentary rock formed from the chemical interaction of limestone and hot springs.
- **Chalk** - a **fine-grained** sedimentary rock made of the fossil remains of diatoms, microscopic animals. It is also called a biochemical sedimentary rock.
- **Limestone** - a **fine to coarse-grained** sedimentary rock made of lime mud. It is also called a chemical sedimentary rock. If it contains fossils, it is then called a biochemical sedimentary rock.
- **Coquina** - a **coarse-grained** sedimentary rock made of fossil shells cemented together with calcium carbonate. It is also called a biochemical sedimentary rock.
- **Bituminous and lignite coal** - sedimentary rocks made of the compressed remains of plants. These are also called biochemical sedimentary rocks.

There will be more rocks you will discover that also fit into the above classification. But for now the rocks your child will be finding the most of will be the above rocks. These are included in your kit. And there you have it.

What about the origin of the rocks?
We will discuss the Biblical view of the origin of these various rock types as we go forward in our study.

You as the teacher should try to demonstrate the importance of studying and collecting the rocks in connection to the creation and flood. These are the two most significant geologic events in our history and studying the rocks can bear this out. You can also help them to think critically and to recognize biases that are unhealthy and often go by without challenge.

Materials provided in your kit:

1. Rock samples for the activities, mineral samples for the activities and a set of practice rocks.
2. Several laminated cards and a tube of E6000 special cement for creating display or reference cards for school projects or future use.
3. Pocket magnifier for close up viewing.
4. Instructional book
5. Rock ID Field Guide

Instructions

1. Familiarize yourself with all the materials in your kit before you begin, so that you know where everything is.
2. Learn the rock-forming minerals before proceeding on to rock identification. Also learn the elements that make up the minerals. This will give you an idea of why certain minerals are light colored and dark colored.
3. Concentrate first on learning to recognize the difference between coarse-grained and fine-grained rocks in Lesson 1.
4. Learn to recognize which minerals are involved in making light colored and dark colored rocks.
5. Learn the geology terms as you go. It is like a language. Before you can speak it, you have to learn a few basics first.

6. The following advice is directed toward those who want to know the origins of rocks and minerals: become familiar with the Biblical framework for interpreting the geology of the earth. This will help you wade through all the books about the age of the earth that tend to produce confusion and challenge to a life of faith.

If I can help you in any way with your project, please email me at northwestexpedition@msn.com or phone me at 425-488-6848.

Lesson 1 – Let's Begin...at the Beginning

How were rocks formed? Do they form now? What is the geology of rocks? Why do they hold kids' fascination? These are exciting questions and that is what we will be exploring in your Rock Identification Made Easy kit.

First, let's talk about the origin of rocks. Because I believe that the Scriptures are reliable historical documents, I will use them as my framework for exploring the rocks. Now, let's be clear. The Bible does not tell us directly about rocks and their origin. But, we can get a glimpse into their origin by using a little common sense. Let's begin.

Genesis 1:1 tells us that God created the space and the earth on day one of creation. There were no massive volcanoes erupting poisonous gasses into the air and no erupting magma onto the boiling lava surface of the early earth. This is the picture that secular geologists paint. But Genesis tells us that in fact on the very first day of creation the Spirit of God was moving over the surface of the ***water***. So, following our Scriptural framework we can say:

1) There was the empty space in which the earth was placed
2) The early earth was covered in water
3) The word 'create' is only used in the Scripture of an ability that God possesses. It means that He brought the space and the earth into being from nothing. The early church fathers called it 'Ex nihilo' - out of nothing.

Peter, one of the apostles of Christ, gave us this same message when he wrote, "...by the word of God the heavens existed long ago and the earth was formed out of water and by water, through which the world at that time was destroyed, being flooded with water..." (2 Peter 3:5-6).

Because water and the earth existed from the beginning, the elements must have also existed from the beginning, because all matter is made of elements. Secular geologists teach that the earth's water was formed out of millions of years of repeated volcanic eruptions. So, either they are wrong or the Scripture is wrong. I like to think that God's Bible has it right!

An element is the most basic of substances that form our earth and space. Elements consist of precisely arranged protons, neutrons and electrons that give substance to all the rocks on our earth. We call these arrangements atoms, which means 'cannot be broken down.' Although we cannot see the atom, scientists usually picture it this way:

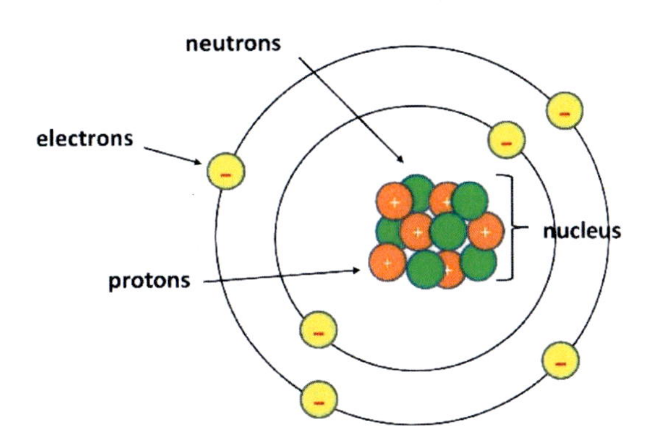

All the elements are made of these precisely arranged atoms. Rocks and minerals make up the earth's crust. Since we are concerned primarily in this book with the rocks of the earth, let's take a look at the ***most abundant*** elements in the earth's crust. The earth's crust is made up of just the right amount of atoms just like the one above but with different numbers of neutrons, protons and electrons. These are called **elements**.

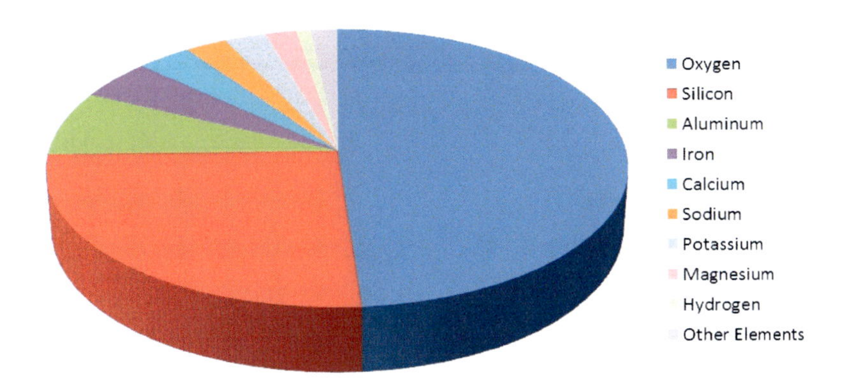

These elements have been combined in special ways to produce **minerals**. Most of the rocks on earth are made of just 12 minerals we call **rock-forming minerals**.

So, **atoms** have been combined in very special ways to produce **elements**. Elements have been combined in very special ways to produce **minerals**. And minerals have been combined in very special ways to produce **rocks**.

Next, let's take a look at the chemical structure of the rock-forming minerals and see how elements have been combined to form their structures.

What is a mineral?
A mineral is a precisely arranged group of elements that exists naturally. It can have a special crystalline shape or it can have what we call a cryptocrystalline shape. That is, the structure of the mineral is not visible to the naked eye. The word cryptocrystalline means 'hidden crystal.' Let's look at one of the most abundant minerals on earth - quartz. In its crystalline form, it always has 6 sides. And it can be in a wide variety of colors. Quartz is found in just about every rock you will find, even volcanic rocks!

Crystalline quartz

Cryptocrystalline quartz

The rock-forming minerals
We call certain minerals the rock-forming minerals because they make up most of the rocks. Further these rock-forming minerals are divided into 6 light-colored and 6 dark-colored minerals. That makes things easy doesn't it? It will be important to learn these 12 minerals because most of the identification of your rocks will be found right here. So, here are **the 6 light-colored rock-forming minerals**:

The 6 light-colored rock-forming minerals

These 6 light-colored rock-forming minerals are from left to right, top row, *quartz, muscovite mica;* middle row, *jasper (a type of quartz colored by iron), potassium feldspar;* bottom row, left to right, *sodium feldspar, and calcite (primarily the mineral found in sedimentary rocks).* **These minerals make rocks light-colored.**

The 6 dark-colored rock-forming minerals are:

The six dark-colored rock-forming minerals are from left to right, top row, *calcium feldspar, biotite mica;* middle row, *pyroxene, amphibole (hornblende);* bottom row, left to right, *magnetite (iron) olivine.* **These minerals make rocks dark-colored.**

Let's take a bit closer look at the rock-forming minerals. In order to become proficient at identifying rocks, we have to be familiar with the rock-forming minerals.

The rock-forming minerals and their elements:

Quartz and jasper - a very hard and abundant mineral made up of the elements silicon and oxygen. Varieties of quartz include agate, chalcedony, chert, flint, opal, and rock crystal. Quartz belongs to the family known as chalcedony. Another type of chalcedony (quartz) is **jasper**, abundant in all types of rocks. Its most common color is red that comes from its iron content.

Feldspar - the most abundant mineral on earth (about 60% of the earth's crust), feldspar, is made up of the elements silicon, oxygen, potassium, aluminum, sodium and calcium, depending on the specific feldspar and the amounts of each of the elements above. For example **potassium feldspar**, or K-feldspar (**orthoclase feldspar**), is made up mostly of silicon, oxygen and potassium. It is light-colored and tends to be pinkish or salmon in color. Granite contains an abundance of potassium feldspar. **Sodium feldspar**, or Na-feldspar (**plagioclase feldspar**), is made up of silicon, oxygen, aluminum and sodium and tends to be white. Another type of feldspar, **calcium feldspar**, or Ca-feldspar (also **plagioclase feldspar**), is made up of the above but also an abundance of calcium. It tends to be darker in color, a representative of which is common labradorite.

Biotite mica; muscovite mica - **Biotite** is a black, iron-rich form of mica. It looks like sheets of black shiny plastic wrap. **Muscovite** (white) mica has the same appearance, but is light colored. What makes them different? Biotite has magnesium and iron in its makeup and so is dark. Muscovite lacks these elements.

Olivine - named for its typical olive color. It is also called chrysolite. Olivine is a mineral that is made of silicon, oxygen, magnesium and iron. The gem variety of olivine is called peridot. Typically olivine is associated with the darker volcanic and plutonic rocks, such as basalt and gabbro.

Pyroxene (the word means, 'fire stranger;' these black minerals were first studied in volcanic rocks where they have the appearance of oddities within the rock; they stand out as isolated small black crystals, hence the name, 'fire stranger.') Pyroxene is a dark blocky mineral containing silicon, oxygen, aluminum, calcium, sodium, magnesium and iron. A common type of pyroxene is **augite**. It is generally in a blocky form. It can look a lot like amphibole or hornblende. But it does not occur in needles that characterize amphibole or hornblende.

Amphibole (the word means, 'ambiguous') - a dark mineral containing oxygen, silicon, sodium, aluminum, magnesium, and iron. A common type of amphibole is **hornblende**. It is common in many igneous and metamorphic rocks. It occurs along with biotite in many cases and it is easy to confuse the two. Biotite is black, occurs in sheets and is shiny. Hornblende is most often black also (but can be dark green or dark brown) and generally occurs in elongated needles as opposed to sheets.

Magnetite (a form of iron) or lodestone, is a typical iron oxide but has magnetic properties. It is made of the elements iron and oxygen. Iron is common in darker igneous rocks, such as basalt. In fact the black sand which forms the beach of Spirit Lake from the volcano, Mt. Bachelor in Oregon is iron basaltic sand. It is fascinating, as this sand will respond to magnets because of its iron content.

Calcite - a mineral made up of the elements calcium, carbon and oxygen. Calcite comes in a variety of colors, is rather soft and is found in abundance in sedimentary rocks.

One of the biggest keys in identifying rocks is their dark or light color. Certain rocks are only light colored and certain rocks are only dark colored. Knowing this will greatly aid your rock identification! *So, remember this tip*: **Light colored rock-forming minerals form light colored rocks. Dark colored rock-forming minerals form dark colored rocks.** Wasn't that easy?

Activity: *You probably have all kinds of rocks in your garage or back yard. And these are your own very special collection of treasures! See if you can sort your rock collection into light colored and dark colored rocks.*

Activity: *Take out the rock-forming minerals laminated card from your kit and place your rock-forming minerals from your kit into the right places.*

I have enclosed a tube of special cement that you can use to glue your minerals into place, if you like. Don't do this right away though, as you may want to do this exercise many times until you have learned the rock-forming minerals. Your card should look something like this when you are done:

The Light Colored Rock-forming Minerals

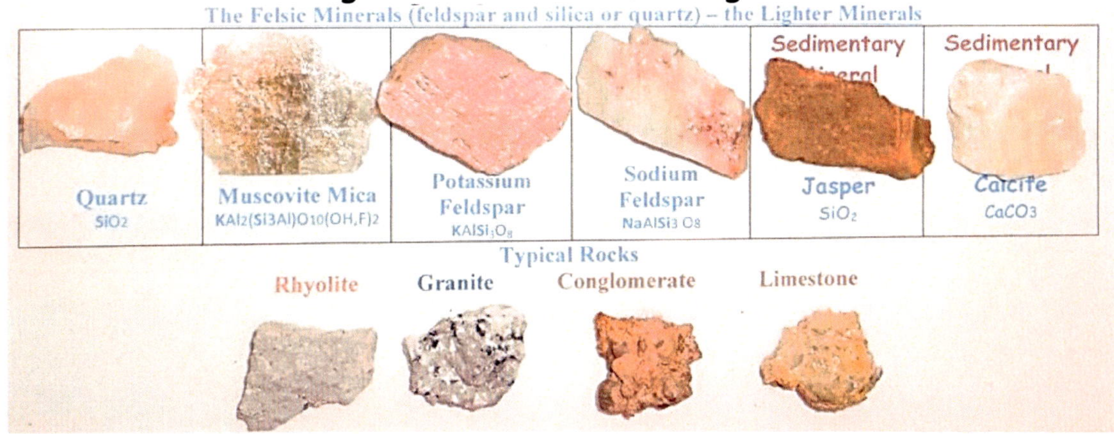

The Dark Colored Rock-forming Minerals

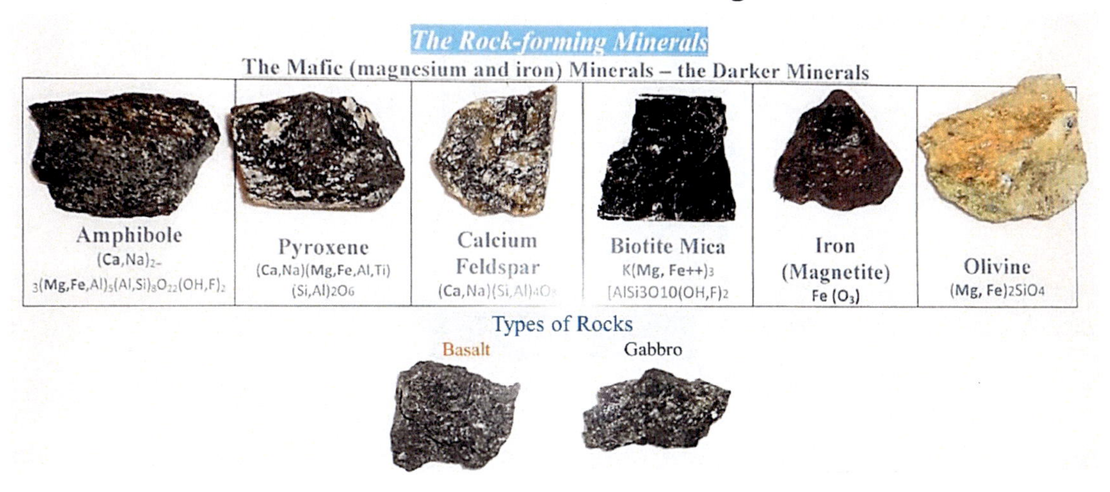

The Elements

Notice on your laminated card that quartz is made of the elements silicon and oxygen arranged in a very special way. Silicon and oxygen are the most abundant elements in the earth's crust. That is why quartz occurs in just about every rock type there is. Together quartz and the feldspars are found in almost every rock in nature.

Activity: *Go through each one of the rock-forming minerals from your kit and write out the elements found in the rock-forming* *minerals. Learn the predominant elements for the light and dark colored rock-forming minerals. This will be very helpful later.*

Did you notice just how prevalent the elements of the earth's crust are in the rock-forming minerals? Compare the elements listed on your chart of the most abundant elements and the elements that are in the rock-forming minerals.

Let's review:

1. **Elements are made of atoms.**
2. **Minerals are made of elements.**
3. **Rocks are made of minerals.**
4. **Rocks have formed all parts of the earth's crust.**

Two other terms that you need to be familiar with are ***felsic*** and ***mafic***. What on earth do these words mean? *Felsic* stands for feldspar (potassium feldspar) and silica (quartz). When certain rocks are light colored, we say that they are ***felsic***, meaning that they are rich in feldspar and quartz.

Mafic stands for magnesium and iron. When certain rocks are dark colored, we say that they are ***mafic***, meaning that they are rich in magnesium and iron.

You can call the light colored rock-forming minerals ***felsic*** minerals, with the exception of calcite, which is high in calcium carbonate. You can call the dark colored rock-forming minerals ***mafic*** minerals, meaning that they are rich in magnesium and iron.

Quiz: Have somebody hold up each of the rock-forming minerals from your kit and quiz you. Can you name them? How did you do? Repeat this quiz until you have this down cold!

Now, let's apply what you have learned and start identifying the rocks. You will begin this in Lesson 2.

Lesson 2 - The Plutonic Rocks and How to Identify Them

Plutonic rocks were named after the mythical god of the underworld, Pluto. They were given that name because these rocks form the foundation of the earth. The continental crust of the earth is made of the plutonic rocks, primarily granite. Although no one has seen granite forming, we know that God created the earth as a functional planet right at the beginning with water and a foundation. Geologists know that the continents are anchored in granite.

So, what do plutonic rocks look like?

Plutonic rocks are **coarse-grained** rocks. That means that you can see the individual minerals that make up the rock. You may not know what those minerals are, but you can see them. Further, plutonic rocks are divided into light colored plutonic rocks and dark colored plutonic rocks.

1. **The light colored plutonic rocks or *felsic* rocks consist of the granites:**

Examples of Granite from Brazil, Utah and Washington

Activity: Take out the samples of granite in your kit and look closely at them. Describe what you see. Can you identify the light colored rock-forming minerals that make up granite?

If you said quartz, mica, and feldspar, you would be correct.

2. Another light colored plutonic rock that looks a lot like granite is a rock called **pegmatite**. The word pegmatite means 'something fastened together.' The distinguishing characteristic of pegmatites is that the mineral crystals are much larger than in common granite. Everything is bigger in a pegmatite.

Pegmatites from Brazil and Colorado

Activity: Take out the sample of pegmatite in your kit and look closely at it. Describe what you see. Now, look at your own collection of rocks from your yard or garage and see if you have any pegmatites.

3. Now let's look at some of the **dark colored plutonic rocks or *mafic* rocks.** These rocks are the gabbros:

Examples of Gabbro from Montana, Wyoming, Washington and California

Activity: Take out the samples of gabbro in your kit and look closely at them. Describe what you see. Can you identify the dark colored rock-forming minerals that make up gabbro?

If you said calcium feldspar, biotite mica, pyroxene, amphibole and iron, you would be correct.

Now, compare the granite with the gabbro. What do you notice that is different between them?

Gabbro is a name of a city in Italy where gabbro was first studied and typified.

Activity: If you have rocks in your garage or back yard, go get them and see if you can find any coarse-grained rocks; organize them into coarse-grained light colored rocks and coarse-grained dark colored rocks. How did you do?

Granite and gabbro are easy to identify. But there is another kind of plutonic rock that is not so easy to identify. These rocks are called *intermediate rocks* because they are in the middle. Let's look at some of the **intermediate plutonic rocks**:

Granodiorite from Butte, Montana and the Cascades of Washington

- **Granodiorite**, as the name implies, is an in-between rock. Grano- means that it is light colored. Diorite means that it tends toward being darker. Some people see these rocks as lighter, while other see them as darker. Why is this? Granodiorite is made up of an *even* mix of the light and dark colored rock-forming minerals. Geologists call these kinds of rocks 'salt and pepper' rocks. The main thing that makes the difference between this intermediate rock and the next one we will look at is that granodiorite *contains some quartz.*

Activity: Take out the samples of granodiorite from your kit and look at them closely. Can you spot the quartz in them? You may need your handy magnifier from your kit to complete this activity.

- **Diorite** is another intermediate rock also called a 'salt and pepper' rock. Diorite differs from granodiorite in that it *contains no quartz,* just an even mix of light and dark colored rock-forming minerals.

Diorite from California

Activity: Take out the sample of diorite from your kit and look at it closely. Can you tell the difference between it and the granodiorite? You may need your handy magnifier from your kit to complete this activity. Look for the quartz in the granodiorite. Don't get frustrated if you cannot quickly identify diorite and granodiorite. It takes practice. I can tell you that the intermediate plutonic rocks are the most difficult to identify.

Activity: If you have rocks in your garage or back yard, go get them and see if you can find any coarse grained intermediate colored plutonic rocks. How did you do?

Simple Rock Classification Chart

Plutonic Rocks: Coarse-Grained Rocks

Rock Name	Light Colored Composition	Intermediate Composition	Dark Colored Composition
Granite	Quartz, K-feldspar, amphibole, biotite		
Granodiorite		Quartz, amphibole, Na-feldspar, biotite	
Diorite		Same, but no quartz	
Gabbro			Ca-feldspar, pyroxene, biotite, amphibole, olivine, iron
Pegmatite (Large Crystals)	Quartz, K-feldspar, amphibole, biotite		

Rounded or tumbled coarse-grained plutonic rocks from Whidbey Island, Washington; remember, you can see the mineral crystals in a plutonic rock. That is why geologists call them **coarse-grained** rocks. You can find these in gravel pits, by rivers and on beaches.

Quiz - the plutonic rocks:

1. Plutonic rocks are also called ______________ - ______________ rocks.
2. ____________ colored rocks are formed from the dark colored ___________________
3. The light colored ___________ are formed from the ___________ colored ___________________
4. In coarse-grained rocks, the minerals ________ be _____________

All the answers to these questions can be found in the text. When you feel you have been able to consistently identify the plutonic rocks, you can move on to the next lesson - Volcanic Rocks.

Lesson 3 – The Volcanic Lava Rocks and How to Identify Them

Volcanic rocks as the name implies, come from volcanoes. Volcanic rocks are the products of volcanic eruptions. Volcanic rocks are called **fine-grained** rocks because normally you cannot see the individual minerals in the rocks, just the general color. And like plutonic rocks, volcanic rocks are divided into dark colored volcanic rocks, light colored volcanic rocks and intermediate colored volcanic rocks. And also like plutonic rocks, it is the dark colored or light colored rock-forming minerals that make up volcanic rocks.

1. Let's take a look at the **light colored volcanic rocks**:
   - **Rhyolite** is a light colored volcanic rock rich in the light colored rock-forming minerals like quartz, and potassium feldspar. It is quite a beautiful rock as you can tell by the examples pictured below. The word rhyolite means 'to flow.' The beautiful patterns that you see in some of the pictures below are actually flow patterns. The flow patterns of the moving rhyolite lava were preserved creating the very interesting color patterns you see below.

Rhyolite from Utah

Examples of the light colored volcanic rock rhyolite from California and Yellowstone

Activity: Take out the samples of rhyolite from your kit and look at them closely. Do any of them have visible flow patterns? How would you describe the colors of your rhyolite samples?

Activity: If you have rocks in your garage or in the backyard, go get them and see if you have any rhyolite lava rocks.

- **The glassy rhyolite lavas**

Obsidian is dark volcanic rock made of pure quartz. So, even though it is dark in color, it is grouped with the lighter-colored lavas. And obsidian comes in all sorts of colors depending on the amount of iron in its makeup. The iron content is what gives obsidian its darker color. Otherwise, it would be just like looking through a window.

Examples of obsidian from California and Mexico

Activity: If you have rocks in your garage or in the backyard, go get them and see if you have any obsidian.

2. Let's take a look at the **dark colored volcanic rocks**:
   - **Basalt** is dark gray to black fine-grained volcanic rock. It is dark in color because it is made up of the dark colored rock-forming minerals. The word basalt means 'hard.' In the Northwest USA, basalt is abundant from all the numerous lava flows and it is used as landscaping rock and road-surfacing rock because it is cheap and holds up to wear and tear. The most common form of basalt is aphanitic basalt. The word aphanitic means 'cannot be seen.' It just appears to be a plain black rock.

Black aphanitic basalt, Western Washington

Basalt lava flow on the Island of Hawaii

Basalt lava moves rather quickly, as it does not have much quartz in its makeup. It is also the hottest lava produced. Basalt is the lava that forms lava tubes. The crust on the outside of a flow cools and hardens while the lava on the inside keeps flowing, producing a cave tunnel. These are pitch black on the inside and using a flashlight is a must!

Basalt is what forms the lava tubes found all over the world. These two pictures are of a basalt lava tube on the Big Island of Hawaii. The first picture is of the inside of the tube. The second is the entrance to the lava tube.

(Author standing at the entrance of a basalt lava tube on the Big Island of Hawaii)

Basalt A'a lava from Hawaii; A'a lava is spikey (or clinkery); the picture on the right above is from the A'a lava flows from near Kona, Hawaii

Basalt is what forms the black sand beaches in Hawaii too.
Author with a sea turtle; these turtles frequent these black beaches;

- Basalt can also be smooth. The smooth basalt is called pahoehoe (puh-OI-OI). It is a ropy kind of lava like the picture below.

Pahoehoe from Craters of the Moon, Idaho

Pahoehoe lava from the Big Island of Hawaii

- Basalt can also display a lot of holes or gas pockets as this picture of basalt from Yellowstone Park shows. We call this fine-grained volcanic rock, **vesicular basalt**, after the holes or vesicles.

- Another form of basalt is called **scoria.** It is from the Greek word meaning, *excrement*. Scoria was originally identified as the waste or refuse from melting metals. Scoria is not as dense as vesicular basalt and has an abundance of holes or vesicles. It is best described as foamy or frothy basalt lava.

Scoria; many more vesicles than vesicular basalt - Three Sisters, Oregon

Other types of basalt that you may find are:

- **Basalt porphyry** with large visible feldspar crystals; with olivine (large visible olivine crystals); with visible sodium feldspar (white) needle-like crystals
- **Basalt glass;** much more dull than obsidian

Basalt glass from California called tachyllite; it is rare.

Activity: Take out the samples of basalt from your kit and look at them closely. Do any of them exhibit porphyry?

Activity: If you have rocks in your garage or in the backyard, go get them and see if you have any basalt lava rocks.

3. The intermediate volcanic rocks

Like the plutonic rocks, volcanic rocks also appear as intermediate rocks. These are generally shades of gray.

- **Andesite** – a light to dark gray fine-grained volcanic rock that most often contains visible tiny feldspar or pyroxene crystals:

Examples of andesite from Western Washington
The bottom sample is more of a porphyry - meaning that the crystals are visible, though small, and they are in a fine-grained matrix.

- **Andesite porphyry** - andesite with large visible mineral crystals of usually feldspar or pyroxene

This is an example of andesite porphyry from Whidbey Island, Washington - visible blocky feldspar crystals and a few black pyroxene crystals. You might think that this is a plutonic rock because you can see the mineral crystals. The difference is the large mineral crystals are in a fine-grained matrix.

Tumbled variety of volcanic porphyritic rocks from Whidbey Island, Washington; the white blocky minerals are feldspar.

- **Dacite** - an intermediate fine-grained volcanic rock that is generally in shades of gray. The main difference between dacite and andesite is that it has more quartz in its makeup and so tends to flow at a slower pace than andesite lava. Because of this, it tends to preserve its flow patterns like rhyolite does.

This is an example of dacite from near Glass Mountain, California. Notice the darker lines in this sample of dacite. These are the flow patterns.

This sample of dacite from California has all kinds of visible feldspar and hornblende crystals in its makeup. Again, some of the mineral crystals are somewhat visible but they're in a fine-grained matrix.

Activity: Take out the samples of dacite and andesite from your kit and look at them closely. Do any of them exhibit porphyry? Do any of them exhibit flow bands?

Activity: If you have rocks in your garage or in the backyard, go get them and see if you have any andesitic or dacitic lava rocks.

Lesson 4 – The Pyroclastic Volcanic Rocks

The word pyroclastic means 'fire broken.' It has to do with its makeup. When an andesitic or rhyolitic volcano first erupts, it explodes with a tremendous amount of very hot gas (1,450° F) both up into the atmosphere and also as rolling clouds of gas and rock, moving at speeds of up to 450 miles per hour. It contains steam, glass particles and bits and pieces of blown-out volcanic rock and ash, hence, the name 'pyroclastic.' The small particles of volcanic rock are called ash. This fast-moving ash flow acts as a sandblaster, stripping everything in its path. When it comes to rest, it cools, fusing together into a hard rock called tuff.

The basalt equivalents, although not as explosive, are called 'bombs' and cinders. Keep your eye open for these things in volcanic landscapes. They are often overlooked.

- **Ash**

Ash from Mt. St. Helens; it is more coarse close to the volcano

Ash from Mt. St. Helens; it is more fine several miles from the volcano

- **Pumice**

Pumice from Mt. St. Helens; mostly a frothy rock and mostly empty space. It therefore usually floats on water

- **Ash-fall tuff**

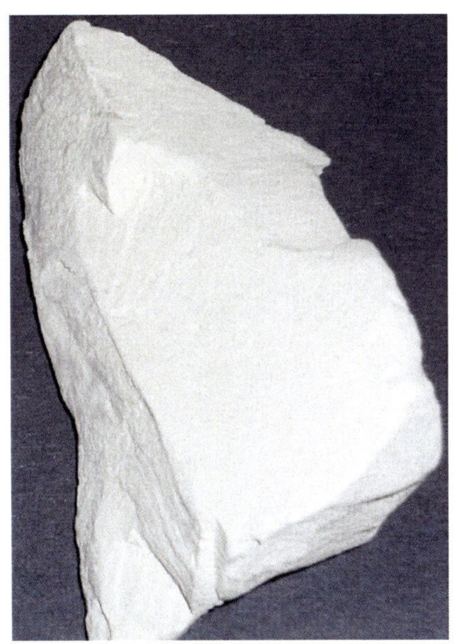

Tuff – a very fine-grained tuff: found in Klamath, Oregon. Erupted from Mt. Mazama, Oregon

- **Welded Tuff** - fused ash, minerals and bits of volcanic rocks into a hard rock

Found in Burns, Oregon, erupted from Mt. Mazama

San Bernardino Mountains, California

The Dalles, Oregon, erupted from Mt. Hood

Examples of tuff varieties, erupted from the Yellowstone caldera

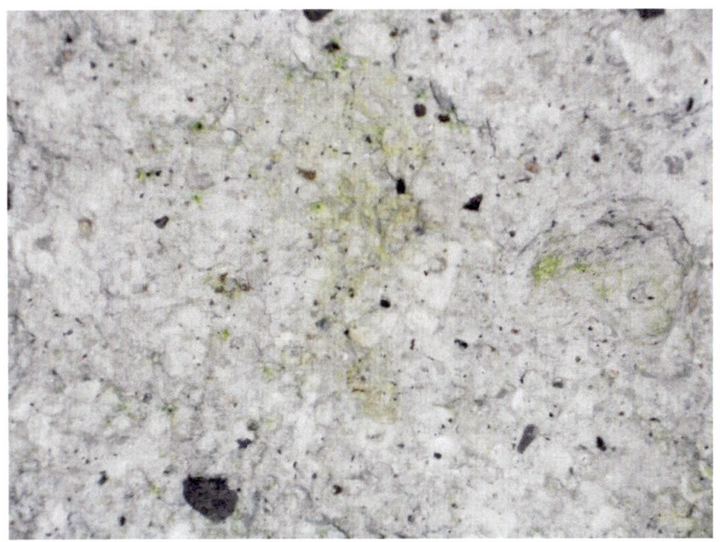

Tuff: close up showing the bits and pieces of rocks and minerals in a light-colored matrix of quartz and feldspar

- **Bombs** - fragments of hot lava ejected from mild basalt eruptions, freezing into various interesting shapes as it flies through the air

Basalt bombs from El Malpais, New Mexico

- **Cinders** - bits and pieces produced from showers of mild basalt eruptions

Basalt cinders from Craters of the Moon, Idaho

Activity: Take out the samples of pyroclastic rocks from your kit and look at them closely. Try floating the pumice in water. What happened? Now try the scoria. What happened?

Activity: If you have rocks in your garage or in the backyard, go get them and see if you have any pyroclastic rocks.

This is a variety tumbled volcanic rocks from Whidbey Island, Washington.
How many different volcanic rocks can you identify?

Simple Rock Classification Chart

Volcanic Rocks - Fine-Grained Rocks

Rock Name	Light Colored Composition	Intermediate Composition	Dark Colored Composition
Rhyolite	Quartz, K-feldspar, amphibole, biotite		
Dacite		Quartz, amphibole, Na-feldspar, biotite	
Andesite		Same, but less quartz	
Basalt			Ca-feldspar, pyroxene, biotite, amphibole, olivine, iron; Very little quartz
Pyroclastic Rocks	Quartz, K-feldspar, amphibole, biotite	Quartz, K-feldspar, amphibole, biotite	
Basalt Bombs			Ca-feldspar, pyroxene, biotite, amphibole, olivine, iron; Very little quartz

Quiz – the volcanic rocks:

1. Volcanic rocks are also called ______________ - ______________ rocks.
2. ____________ colored rocks are formed from the dark colored __________________
3. The light colored ___________ are formed from the ___________ colored __________________
4. In fine-grained rocks, the minerals __________________ be _____________
5. A gray volcanic rock might indicate that you have an _____________________________ volcanic rock
6. The word pyroclastic means ____________ __________________
7. Obsidian is technically what kind of lava?

When you feel you have been able to consistently identify the volcanic rocks, you can move on to the next lesson - Metamorphic Rocks.

Lesson 5 – The Metamorphic Rocks

The word metamorphic means 'changed form.' Secular geologists think that these rocks were changed from original parent rocks deep underground over hundreds of millions of years, through heat and pressure. Now, geologists have never seen metamorphic rocks forming. And of course, if the Scriptures mean anything, they do not teach this kind of earth history time frame. So, there must be a different explanation. They do appear to have been changed. But the process through which they were changed could have been accomplished through the tectonic processes of the Genesis Flood, making it a fairly rapid process.

Metamorphic rocks are divided according to what geologists call 'foliated' meaning that they are either banded, usually in alternating light and dark colored minerals or layered, and 'non-foliated.' meaning that they are crystallized; they appear to have had their minerals fused together in some way.

1. The foliated metamorphic rocks

These rocks appear to have undergone metamorphosis of a parent rock, possibly through heat and pressure.

Metamorphic foliated rock	Suspected parent rock/mineral
Gneiss - banding	Granite, diorite or gabbro
Slate - layering	Shale
Schist - layering	Clay/mica
Phyllite - layering	Slate/mica

The foliated metamorphic rocks in your kit are:

- ***Gneiss***

Examples of gneiss from the Yellowstone area – supposedly metamorphosed diorite, granite, and diorite

- **Slate**

Examples of slate: supposedly metamorphosed shale; slate can be red, black, gray or even brown. From California.

- **Schist**

Example of schist from the Cascades of Washington – supposedly metamorphosed clay with mica; looks like a rock with lots of glitter glued to it

- **Phyllite**

Example of phyllite, from California: supposedly metamorphosed slate with the addition of mica; it has a very shiny, smooth surface

2. The non-foliated metamorphic rocks

These rocks appear to have undergone crystallization of a parent rock, possibly through heat and pressure.

Metamorphic non-foliated rock	Suspected parent rock/mineral
Quartzite	Sandstone/quartz
Marble	Limestone
Serpentinite	Clay/serpentine

The nonfoliated metamorphic rocks in your kit are:

- **Quartzite**

Examples of quartzite from Brazil and Montana; it has a sugary appearance and is a very hard rock; the quartzite on the right still preserves the iron oxide banding from the sandstone from which it came

Tumbled quartzite from Whidbey Island, Washington; quartzite is extremely hard and therefore will tumble to a brilliant shine.

- **Marble**

Here are two examples of marble from Canada and Washington; one variety has a sugar cube like appearance and will react to hydrochloric acid by fizzing. The fizzing indicates the presence of calcium carbonate, which would indicate that it is likely metamorphosed limestone, which also contains calcium carbonate, but is a sedimentary rock. (We'll discuss limestone later.) Another variety, called Picasso Marble, is characterized by non-aligned banding, characteristic of Picasso's paintings. Marble comes in an almost unlimited variety or colors and patterns

- **Serpentinite**

Example of Serpentinite from the west side of Yosemite National Park, California

Serpentinite from Utah

Two varieties of serpentinite; they both display the classic serpent green color of the mineral serpentine

Activity: Take out the samples of metamorphic rocks from your kit and look at them closely. Do any of them exhibit banding? Do any of them exhibit layering?

Activity: If you have rocks in your garage or in the backyard, go get them and see if you have any metamorphic rocks.

A variety of tumbled metamorphic rocks from Whidbey Island, Washington. How many can you identify?

Quiz – the metamorphic rocks:

1. Metamorphic rocks are divided into two groups:
   a) ______________________ metamorphic rocks (banded or layered)

   b) ______________________ metamorphic rocks (crystalline)
2. Metamorphic rocks are thought to be metamorphosed _________________ rocks (came from another rock)
3. Fill in the following chart:

Metamorphic rock	Primary mineral(s)	Parent rock
Gneiss (dark colored)		
Gneiss (light colored)		
Slate		
Schist		
Quartzite		
Marble		
Serpentinite		

When you feel you have been able to consistently identify the metamorphic rocks, you can move on to the next lesson – Sedimentary Rocks.

Lesson 6 - The Sedimentary Rocks

Sedimentary rocks are those that have been laid down by or in water and mud. Seventy per cent of the earth's surface is covered in sedimentary rocks. This means that there have been an awful lot of muddy sediments in earth's past. The Genesis Flood is the only reasonable geologic event that can account for all this sediment.

Geologists divide sedimentary rocks into 3 groups based on their composition.

1. **The clastic sedimentary rocks**; the word 'clastic' is a Greek word that means 'broken.' So clastic sedimentary rocks are those rocks that have been put together with clasts of broken bits and pieces of other rock and minerals and cemented together with silica, calcite or iron oxide. Clastic sedimentary rocks can contain significant numbers of fossils.
2. **The chemical sedimentary rocks**; Chemical sedimentary rocks have been produced by some sort of chemical reaction within the sediments plus water.
3. **The biochemical sedimentary rocks**; These are rocks that have been produced through some sort of chemical reaction involving muddy waters and sediments plus the remains of once-living things.

Let's look at each one of these.

1. **The clastic sedimentary rocks**; the clastic sedimentary rocks are grouped according to the size of the clasts they contain; all the way from very fine to clasts the size of boulders. They are bound together with silica (quartz), calcite or iron oxide. Below is a handy chart to help you group the sedimentary rocks according to clast size. The chart organizes sedimentary clastic rocks according to clast size from extra fine to coarse. Not all the rocks on this chart are in your kit, but the chart does serve as a nice reference chart for future use.

Clastic Sedimentary Rocks

"clast" means "broken" - broken bits and pieces of rock cemented together

Sedimentary Rocks are Rocks Laid Down by Water and Mud

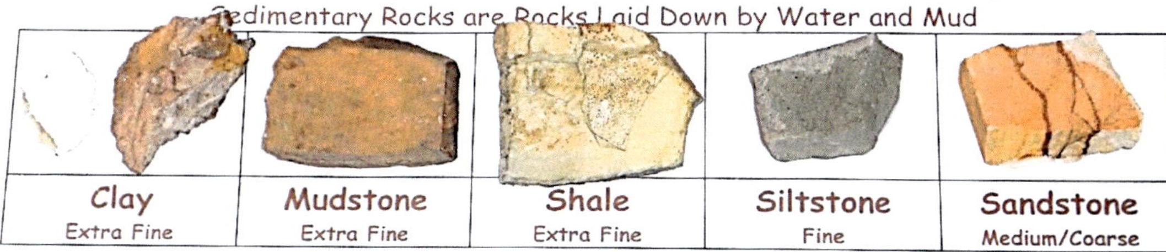

Clay	Mudstone	Shale	Siltstone	Sandstone
Extra Fine	Extra Fine	Extra Fine	Fine	Medium/Coarse

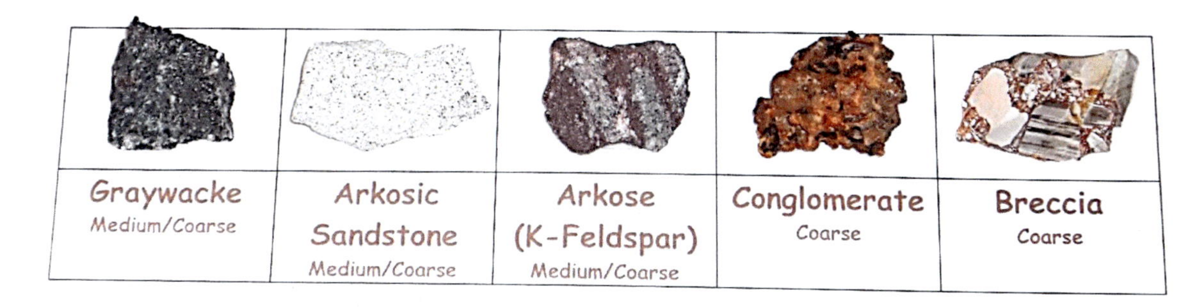

Graywacke	Arkosic Sandstone	Arkose (K-Feldspar)	Conglomerate	Breccia
Medium/Coarse	Medium/Coarse	Medium/Coarse	Coarse	Coarse

The clastic sedimentary rocks in your kit are:

- **Shale** is made of very fine or extra fine clasts of clay. It is also called mudstone or claystone. It can be a variety of colors depending on the elements that have gone into its color. For example red shale has a lot of iron in its composition. The clay clasts have hardened into a layered sedimentary rock. Black shale contains carbon.

Examples of shale from California and Montana

- **Siltstone** is made of fine or very small clasts of sedimentary particles that are larger than clay, but smaller than sand. Magnification is needed to see the clasts. It can be a variety of colors.

Example of siltstone from California

- **Sandstone** is made of coarse clasts of the mineral quartz, usually as rounded clasts, cemented together by silica or calcite. The clasts are easily seen with the naked eye. Depending on the amount of iron oxide in the composition of the sandstone, it can be light colored or very red.

Examples of sandstone, Utah

- **Conglomerate** is made of *rounded* clasts or pebbles cemented together by silica or calcite. The pebbles can be quartz or jasper or even volcanic rock. When the conglomerate is made of volcanic rock, we call it agglomerate.

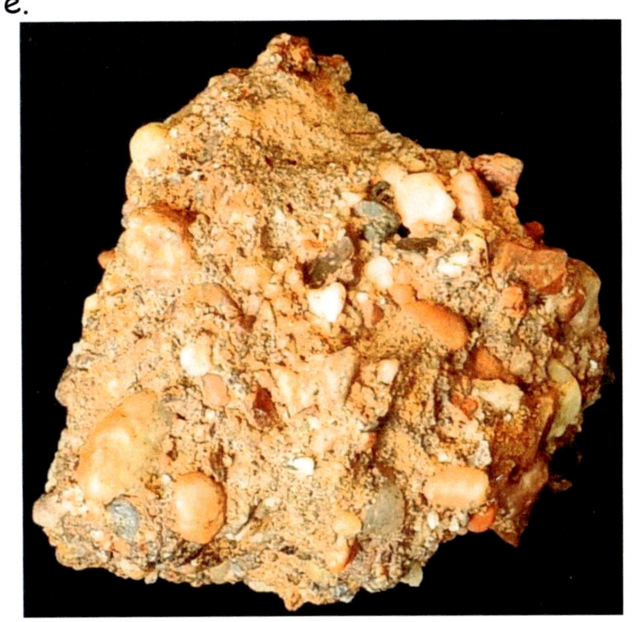

Example of a conglomerate from Texas

Example of agglomerate from Olympic Peninsula, Washington

- **Breccia** is made of *angular* clasts or pebbles cemented together by silica or calcite. The pebbles can be quartz or jasper or even volcanic rock. When the breccia is made of volcanic rock, we call it volcanic breccia.

Examples of sedimentary breccia from Utah

Example of volcanic basalt breccia cemented together by the volcanic mudflows of Yellowstone National Park

Example of volcanic rhyolite breccia cemented together by volcanic mudflows from Nevada

2. **The chemical sedimentary rocks** are rocks that have formed through some sort of chemical process. Below is a handy chart to help you group the chemical sedimentary rocks according to chemical and biochemical classification. Not all the rocks on this chart are in your kit, but the chart does serve as a nice reference chart for future use.

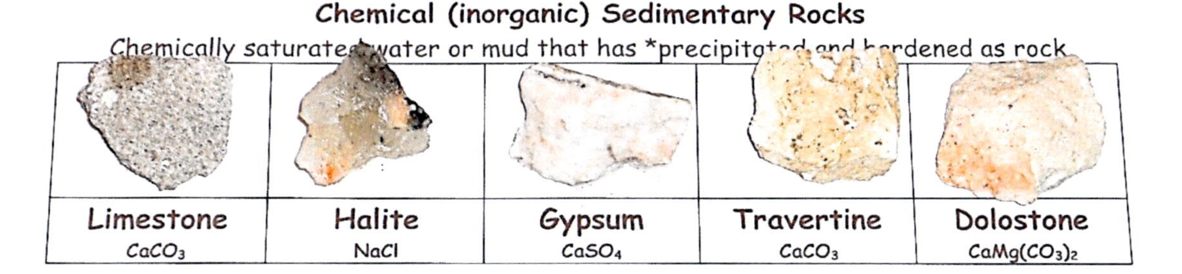

Chemical (inorganic) Sedimentary Rocks

Chemically saturated water or mud that has *precipitated and hardened as rock

Limestone	Halite	Gypsum	Travertine	Dolostone
$CaCO_3$	NaCl	$CaSO_4$	$CaCO_3$	$CaMg(CO_3)_2$

Biochemical (bioclastic or organic) Sedimentary Rocks

Chemical reaction involving organic material that has turned it to rock

Bituminous Coal	Chalk	Chert, Flint	Fossil Limestone	Coquina
Plant matter	Diatoms in $CaCO_3$	Diatoms in SiO_2	Marine fossils in $CaCO_3$	Marine fossils in $CaCO_3$

*Precipitation is a chemical reaction or process that causes a solid to form in/from chemically saturated solutions. It is not the same thing as evaporation, which is an observed slow process that simply leaves a solid residue as the liquid evaporates.

The chemical sedimentary rocks in your kit are:

- **Travertine** – a chemical sedimentary rock formed from the chemical interaction among limestone, hot or cold springs and carbon dioxide. It can be white, yellowish white, or a combination of intense color banding due to iron oxide. Travertine is sometimes called by its commercial name 'onyx.' But it is not onyx. Strictly speaking, onyx is a type of banded agate. Travertine is used in all sorts of building materials and was used in the construction of the Coliseum in Rome.

Examples of travertine from Gardiner, Montana; a chemical sedimentary rock from New Mexico

- **Limestone** – a chemical sedimentary rock made of lime mud. Limestone is one of the most abundant sedimentary rocks on earth. Limestone can be many different colors. The only real way to tell if you have found limestone is to put a small drop of a strong acid on it. If it is limestone, it will react and fizz. That is because it is made of calcium carbonate or calcite which reacts to acid.

Example of limestone from Texas

3. **The biochemical sedimentary rocks** are rocks that have formed through some sort of chemical process acting on living things. **The biochemical sedimentary rocks in your kit are:**

- **Chalk** - a limestone rock made out of the skeletons of tiny plankton. Geologists agree that these must have been formed in deep water. But here is the amazing thing. The deposits of chalk around the world are enormous. The famous white cliffs of Dover, England are made of this fossil plankton. The Genesis Flood adequately explains such huge amounts of chalk. Large amounts of water with concentrated sediments containing billions of dead things would have characterized the Genesis Flood. This would have included plankton.

Example of chalk, Kansas

- **Fossil limestone** - a limestone made from the shells of marine fossils in a limy mud that has been turned to rock. It is an abundant sedimentary rock all over the earth.

Example of fossil limestone from Texas

- **Coquina** – a type of limestone made exclusively from the broken bits and pieces of marine shells. In fact the word, "coquina" is the Spanish word for shell.

Examples of coquina from Florida: second picture is enlarged to show the fossil shells

- **Lignite and bituminous coal** are made of the compressed remains of plants that have undergone a chemical process of some kind to harden them into rocks. Lignite is a brownish coal made of around 30-35%

carbon and has around a 66% moisture content. It is also known as fossil peat. It is abundant all over the world. Around 45% of Germany's electricity comes from lignite power plants. The main difference between lignite and bituminous is that bituminous has around 60-80% carbon content. Bituminous coal is also abundant all over the world.

Example of lignite coal from North Dakota

Example of bituminous coal from West Virginia

Activity: Take out the samples of sedimentary rocks from your kit and look at them closely. Do any of them exhibit banding? Do any of them exhibit layering? Can you see the difference between the chemical and biochemical sedimentary rocks? Do any of them contain fossils? Can you identify the fossils?

Activity: If you have rocks in your garage or in the backyard, go get them and see if you have any sedimentary rocks.

This is rock identification. It can be fun and rewarding to go rock collecting and not just pick up interesting rocks, but to be able to identify them. Identifying the rocks will also help you appreciate the God's handiwork in creation and in the awesome power of the Flood.

Rock identification takes practice. You will improve with time and experience. Most of all you must continue to review the information in this book otherwise you will lose it. It is like a language. You must continually use it in order to keep it in your mind and be able to use it as you go.

Quiz – the sedimentary rocks:

1. Sedimentary rocks are rocks that have been formed in ___________ and __________
2. Name the three groups of sedimentary rocks:
   a) The ______________ sedimentary rocks (bits and pieces of rocks)
   b) The ______________ sedimentary rocks (limy mud)
   c) The ______________ sedimentary rocks (living things)
3. A sedimentary rock that is extra fine-grained and layered is called __________________
4. A sedimentary rock that is coarse-grained and colored by iron oxide is called ____________________
5. A sedimentary rock that has coarse-grained *irregular* shaped bits and pieces of rock is called ________________

6. A sedimentary rock that has coarse-grained *rounded* bits and pieces of rock is called ________________________
7. The rock made from a chemical process involving limestone and hot springs is called __________________________
8. Rocks such as coal, fossil limestone and coquina are called ______________________ sedimentary rocks.

How many rocks can you identify from this pile of gravel?

Appendix A
Where to Find Rocks

Excellent places to find rocks include:

- Industrial rock quarries
- Gravel beds
- Road cutouts
- Stream beds
- Foothills of mountains
- Along gravel roads

For ideas on building and displaying your collection, refer to the back of the Rock Identification Field Guide that came with your kit for building and displaying your rock collection.

One of the best places for finding rocks is in gravel pits or gravel bars along rivers.

This is an unsorted pile of gravel washed up on the shore of Whidbey Island, Washington. These kinds of places are a treasure trove for the rock hound.

Another great place for collecting rocks is along road cutouts of US Highways. Pull off the road and see the amazing assortment of different rocks there are. Be careful though, as traffic can be heavy along these highways. And remember that except for emergencies, you are not permitted to stop along Interstate Highways.

Road cutout along US Highway 287 outside the west entrance of Yellowstone Park. This gravel consists of rounded cobbles of all kinds of the rock types. The rounding of the stones really tells a story of a massive amount of flooding that probably came from the melting of huge glaciers higher in the Madison Range of mountains nearby. This event would have followed the year-long Genesis Flood during which great numbers of volcanoes would have discharged tremendous amounts of ash into the atmosphere and brought about a type of an ice age.

Appendix B
A Biblical View of Geology

James Hutton, the Father of Modern Geology, viewed the earth and its geology as cyclical. He stated that (from his studies of the earth) he could see no prospect of a beginning or of an end. All continued in the present as it always had done in the past. (This concept is known as uniformitarianism.) Therefore God was no longer needed to explain the history neither of the earth nor of the strange formations and fossils that were found in the rock layers. It was up to man to decipher and explain the history of the earth as he discovered its natural processes. This idea was the product of the Enlightenment. It became known as Deism and is a heresy of Theism.

Deism - the idea that God probably created the earth in the remote indiscernible past; things have been left to run their natural course since then. God is no longer involved nor is it relevant in modern geology. Therefore science must explain everything.

Theism - the idea that at one point in recent history God created the earth and the universe, but continues to be involved in its maintenance. Although there are "natural" laws that govern earth's processes, they are subject to God. Therefore only God can explain everything. He has given His Scriptures so that we might know at least some of His Story - history.

One of the most obvious examples of James Hutton's influence on modern geology is the Rock Cycle.

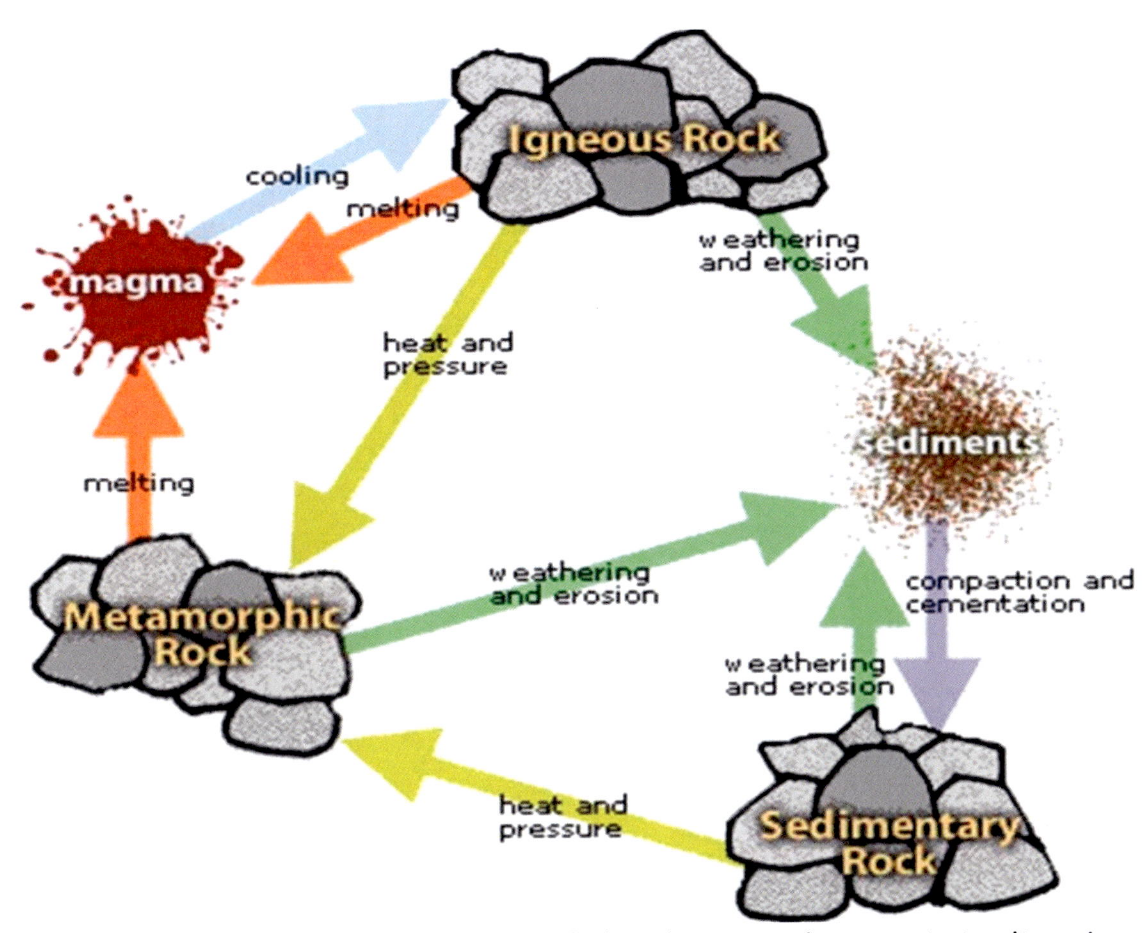

Although it is conceivable that some of the above can happen, it implies that the earth is an eternal state of recycling, just as James Hutton had taught. Even though some of this can be observed today to be happening, the big question is, has it always happened and will it continue to happen just like this?

Where do rocks come from?

Secular geologists tell us that rocks are divided into three basic types, igneous, sedimentary and metamorphic, depending upon how they were formed, and that Plate Tectonics provides an explanation for how rocks are recycled from igneous to sedimentary to metamorphic and back to igneous again.

This represents a typical explanation for the origin of rocks. But here are the questions:

1. Has anyone ever observed the plutonic igneous rocks form?
2. We have observed volcanic rocks forming, but has it always been this way from the beginning?
3. Has anyone ever seen metamorphic rocks forming?
4. Although we have manufactured types of sedimentary rocks, for example, concrete, has this process been going on since the beginning?

5. We also know that there is present shifting of land masses. But does this mean that this action has been going on since the beginning?

Although the above process can be observed to some degree especially with the breakdown of rocks by wind, chemical and water erosion into soil, it can't explain the remote past or the future. If Hutton imagined this same cycle in the late 1700s, it is no wonder that he concluded that he could see no prospect of a beginning or of an end to the earth.

We also have observed experimentally rapid and slow cooling of supersaturated liquids that produce both large and small crystals. But how predictive is this? Granites have been shown radiometrically that they have formed quickly, possibly as a result of the Genesis Flood processes. Furthermore we have not observed granite forming nor any of the other plutonic rocks for that matter. Recent research shows that crystal formation may actually be due more to the environment in which crystals form, not time.

Rock identification involves science; identifying the minerals and types of rocks based on their composition. Most of geology, the study of the earth, involves philosophy: studying the origin of the earth. It should be clear to the student that anything that involves the remote past cannot be tested by science. The origin of the earth is therefore not science, but a collection of codified ideas.

Most people have never studied the Scriptures to see if they have anything to say about earth history. The Bible has generally been considered to be a religious book and therefore has nothing to say about earth history. But that could not be further from the truth. The Bible very succinctly describes earth history in the following accounts:

- **a)** a six day creation about 6,000 years ago
- **b)** a moral fall of man and the entrance of corruption into the world
- **c)** a global flood by which the entire surface of the earth was rearranged through catastrophic upheaval

Biblical geology is all about interpreting the various landforms and the uniqueness of man and life on earth in light of this brief outline or framework. Following is a collection of Scriptures that shed light on earth history. They all follow the historical accounts mentioned above.

A Framework for Biblical Geology

Chapters 1 & 2 of 2 Peter focus on what we have in Christ and what we need to do to experience our salvation here during this present life. Toward the end of chapter 1, Peter reminds us that the apostles did not follow cleverly devised tales but were eyewitnesses. He then tells us to pay attention to the Scriptures as to a light shining in a dark place. In chapter 3 Peter reminds us that in the last days, mocking and denial of the truth will mark the times. This is where we will pick up.

2 Peter chapter 3, 1 "This is now, beloved, the second letter I am writing to you in which I am stirring up your sincere mind by way of reminder, **2** that you should remember the words spoken beforehand by the holy prophets and the commandment of the Lord and Savior spoken by your apostles. **3** Know this first of all, that in the last days mockers will come with their mocking, following after their own lusts, **4** and saying, Where is the promise of His coming? For ever since the fathers fell asleep, all continues just as it was from the beginning of creation." **5** For when they maintain this, it escapes their notice that by the word of God the heavens existed long ago and the earth was formed out of water and by water, **6** through which the world at that time was destroyed, being flooded with water. **7** But by His word the present heavens and earth are being reserved for fire, kept for the Day of Judgment and destruction of ungodly men. **8** But do not let this one fact escape your notice, beloved, that with the Lord one day is like a thousand years, and The Lord is not slow about His promise, as some count slowness, but is patient toward you, not wishing for any to perish but for all to come to repentance. **10** But the day of the Lord will come like a thief, in which the heavens will pass away with a roar and the elements will be destroyed with intense heat, and the earth and its works will be burned up. **11** Since all these things are to be destroyed in this way, what sort of people ought you to be in holy conduct and godliness, **12** looking for and hastening the coming of the day of God, because of which the heavens will be destroyed by burning, and the elements will melt with intense heat! **13** But according to His promise we are looking for new heavens and a new earth, in which righteousness dwells. **14** Therefore, beloved, since you look for these things, be diligent to be found by Him in peace, spotless and blameless, **15** and regard the patience of our Lord as salvation; just as also our beloved brother Paul, according to the wisdom given him, wrote to you, **16** as also in all his letters, speaking in them of these things, in which are some things hard to understand, which the untaught and unstable distort, as they do also the rest of the Scriptures, to their own destruction. **17** You therefore, beloved, knowing this beforehand, be on your guard so that you are not carried away by the error of unprincipled men and fall from your own steadfastness, **18** but grow in the grace and knowledge of our Lord and Savior Jesus Christ. To Him be the glory, both now and to the day of eternity. Amen." Let's look at these Scriptures. Because they outline exactly what is happening in our own culture these very days.

1. **2 Peter 3:4,** The earth and universe had a beginning and it was brought about by the word of God.
2. **2 Peter 3:4,** As God's promise of coming back and ruling the earth would tarry, the prevailing idea would become Deism/Naturalism.
3. **2 Peter 3:4,** The earth would be increasingly viewed in a naturalistic way without a god who is involved in his creation.
4. **2 Peter 3:5,** Naturalism is set at odds w/ supernaturalism:
    - "God said" vs. natural processes
    - The earth was formed out of water and by water vs. a volcanic/fire beginning
    - The original created world was destroyed in a global cataclysm vs. an eternal or timeless recycling of the earth's materials; ex. the Rock Cycle
    - The present heavens and earth will be judged with fire vs. a continuous recycling and renewing of earth's resources
5. **2 Peter 3:7-13,** Biblical geology also incorporates the future of the earth and universe
    - The earth and universe as they exist today are being kept for fire and judgment
    - God exists outside of time and our created universe; He is transcendent
    - God waits to judge in order to give every possible chance to people for repentance
    - The heavens will pass away with a roar
    - The elements will be destroyed w/ intense heat
    - The earth and its works will be burned up
    - Destruction will be final and eternal, not like the Flood
    - There will be a new heavens and a new earth in which righteousness dwells; free from evil and struggle

The subject of Biblical geology in 2 Peter can be arranged according to categories of earth study:

1. The **history** of the earth - the origin of and development of its landforms and its occupants
2. The **present processes** of the earth - erosion, corruption; the effects of sin
3. The **makeup** of the earth - its temporal rocks, soil and its chemistry
4. The **position** of the earth - its relationship to other bodies in the universe and to the spiritual
5. The **future** of the earth - its destruction and the future of mankind

To get a complete worldview we must consider all these elements. Remember that Deism seeks to obliterate any reference to the metaphysical. But to ignore this would be to get a skewed understanding of who we are and what we are doing here. Biblical geology is much more comprehensive than modern geology. The individual references in 2 Peter can be divided according to the following:

- 2 Peter 1:4, 5-7, 8-9, 12, 15, 16, 19, 21 - #2
- 2 Peter 1:11 - #5
- 2 Peter 1:13-14 - #5
- 2 Peter 1:18 - #4
- 2 Peter 1:21 - #4
- 2 Peter 2:2, 13, 18, 20, 21 - #2
- 2 Peter 2:4 - #4
- 2 Peter 2:5 - #1, #4
- 2 Peter 2:6-8 - #4
- 2 Peter 2:9 - #5
- 2 Peter 2:12 - #1, #2, #5
- 2 Peter 2:16 - #4
- 2 Peter 2:17 - #2, #5
- 2 Peter 2:19-20 - #2
- 2 Peter 3:2, 5, 9, 13, 16-18 - #2
- 2 Peter 3:3 - #5
- 2 Peter 3:4-10 - #2 produces the following conclusions:
    1. All continues just as it was from the beginning
    2. The beginning of creation did not involve God
    3. Questioning/doubting God's promises

4. Naturalism becomes the only reasonable alternative
5. The elemental truths of Scripture are missed – the creation came about by the word of God, how the earth was formed, the earth is temporal, the ancient world was flooded, the present heavens and earth are reserved for fire and judgment, God is not idle, God is not bound by time but is patient. This truth is often misunderstood for lack of involvement (Deism). The Day of the Lord will come like a thief. The heavens and the earth will be destroyed w/ intense heat and the earth will be burned up.
6. Because these truths are missed, the Christian will not have a sound worldview concerning this life or of the one to come. Having the correct worldview motivates the following: A belief in God's coming. We will live as if the world will not last. Belief in God's promise of a new heavens and a new earth. We will strive to be blameless as we prepare for our new home. We will embrace the Scriptures and not be motivated by lust. We will be on our guard so that we will not be led astray by the error of unprincipled men. We will grow in the grace and knowledge of Christ.

Earth history is explained in the Scriptures. And it is explained in such a way that is not ambiguous. God has given us enough to have a solid outline or framework for teaching us where we came from, how the rocks got here and where we are going.

Picture credits

Cover Photo: Photo by Heidi Ann Noggle, from the author's collection.

Introduction
Coarse-grained rock: Photo by Patrick Nurre. From the author's personal collection, 6. Fine-grained rock: Photo by Patrick Nurre. From the author's personal collection, 6. Light-colored rock-forming minerals: Photo by Patrick Nurre. From the author's personal collection, 6. Dark-colored rock-forming minerals: Photo by Patrick Nurre. From the author's personal collection, 7.

Lesson 1
Atom: Image by Vicki Nurre, 13. Elements in the Earth's Crust: Image by Vicki Nurre, 13. Crystalline quartz: Photo by Heidi Ann Noggle. From the author's personal collection, 14. Cryptocrystalline quartz: Photo by Heidi Ann Noggle. From the author's personal collection, 14. Six light-colored rock-forming minerals: Photos by Heidi Ann Noggle. From the author's personal collection, 15. Six dark-colored rock-forming minerals (4): Photos by Heidi Ann Noggle. From the author's personal collection, 16. Six dark-colored rock-forming minerals (2): Photos by Vicki Nurre. From the author's personal collection, 19. Light-colored rock-forming minerals on card: Photo by Patrick Nurre. From the author's personal collection, 19.
Dark-colored rock-forming minerals on card: Photo by Patrick Nurre. From the author's personal collection, 19.

Lesson 2
Granites: Photos by Patrick Nurre. From the author's personal collection, 21. Pegmatites: Photos by Heidi Ann Noggle. From the author's personal collection, 22. Gabbro samples: Photos by Patrick Nurre. From the author's personal collection, 23. Granodiorite samples: Photos by Patrick Nurre. From the author's personal collection, 24. Diorite: Photo by Patrick Nurre. From the author's personal collection, 25. Coarse-grained plutonic rocks: Photo by Patrick Nurre. From the author's personal collection, 27.

Lesson 3
Rhyolite lavas: Photos by Patrick Nurre. From the author's personal collection, 28, 29. Obsidian samples: Photos by Patrick Nurre. From the author's personal collection, 30. Basalt: Photo by Patrick Nurre. From the author's personal collection, 31. Basalt lava flow: Photo by Vicki Nurre, 31. Lava tube: Photo by Vicki Nurre, 32. Lava tube: Photo by Vicki Nurre, 32. A'a' lava: Photos by Patrick Nurre. From the author's personal collection, 33. Black sand beach: Photo by Vicki Nurre, 33. Black sand: Photo by Patrick Nurre. From the author's personal collection, 33. Pahoehoe: Photo by Patrick Nurre. From the author's personal collection, 34. Pahoehoe lava: Photo by Patrick Nurre, 34. Vesicular basalt: Photo by Heidi Ann Noggle. From the author's personal collection, 35. Scoria samples: Photos by Patrick Nurre. From the author's personal collection, 35. Basalt porphyry: Photos by Patrick Nurre. From the author's personal collection, 36. Basalt glass: Photo by Patrick Nurre. From the author's personal collection, 36. Andesite lavas: Photos by Patrick Nurre. From the author's personal collection, 37. Andesite porphyry: Photo by Patrick Nurre. From the author's personal collection, 38. Porphyritic rocks: Photo by Patrick Nurre. From the author's personal collection, 38. Dacite: Photo by Patrick Nurre. From the author's personal collection, 39. Dacite: Photo by Patrick Nurre. From the author's personal collection, 39.

Lesson 4
Volcanic ash: Photos by Patrick Nurre. From the author's personal collection, 40. Pumice: Photo by Patrick Nurre. From the author's personal collection, 41. Tuff: Photo by Patrick Nurre. From the author's personal collection, 41. Welded tuff samples: Photos by Patrick Nurre. From the author's personal collection, 42, 43. Bombs: Photo by Patrick Nurre. From the author's personal collection, 43. Cinders: Photo by Vicki Nurre, 44. Volcanic Rocks: Photo by Patrick Nurre. From the author's personal collection, 44.

Lesson 5
Gneiss samples: Photos by Patrick Nurre. From the author's personal collection, 48. Slate: Photo by Patrick Nurre. From the author's personal collection, 48. Schist: Photo by Patrick Nurre. From the author's personal collection, 49. Phyllite: Photo by Patrick Nurre, from the author's personal collection, 49. Quartzite samples: Photos by Patrick Nurre. From the author's personal collection, 50. Tumbled quartzite: Photo by Patrick Nurre. From the author's personal collection, 50. Marble samples: Photos by Patrick Nurre. From the author's personal collection, 51. Serpentinite samples: Photos by Patrick Nurre. From the author's personal collection, 51. Tumbled metamorphic rocks: Photos by Patrick Nurre. From the author's personal collection, 52.

Lesson 6
Clastic sedimentary rocks card: Photo by Patrick Nurre. From the author's personal collection, 55. Shale samples: Photos by Patrick Nurre. From the author's personal collection, 55. Siltstone: Photo by Patrick Nurre. From the author's personal collection, 56. Sandstone samples: Photos by Patrick Nurre. From the author's personal collection, 56. Conglomerate: Photo by Heidi Ann Noggle. From the author's collection, 57. Agglomerate: Photo by Patrick Nurre. From the author's personal collection, 57. Breccia: Photo by Heidi Ann Noggle. From the author's personal collection, 58. Breccia: Photo by Patrick Nurre. From the author's personal collection, 58. Basalt breccia: Photo by Patrick Nurre, 58. Rhyolite breccia: Photo by Patrick Nurre. From the author's personal collection, 58. Chemical sedimentary rocks card: Photo by Patrick Nurre. From the author's personal collection, 59. Biochemical sedimentary rock card: Photo by Patrick Nurre. From the author's personal collection, 59. Travertine samples: Photo by Heidi Ann Noggle. From the author's personal collection, 60. Travertine: Photo by Patrick Nurre. From the author's personal collection, 60. Limestone: Photo by Patrick Nurre. From the author's personal collection, 60. Chalk: Photo by Heidi Ann Noggle. From the author's personal collection, 61. Fossil Limestone: Photo by Heidi Noggle. From the author's personal collection, 61. Coquina: Photos by Patrick Nurre. From the author's personal collection, 62. Lignite coal: Photo by Heidi Ann Noggle. From the author's personal collection, 63. Bituminous coal: Photo by Patrick Nurre. From the author's personal collection, 63. Gravel: Photo by Vicki Nurre, 65.

Appendix A
Photo by Vicki Nurre, 66. Road cut-out: Photo by Patrick Nurre, 67.

Appendix B
Rock Cycle: http://mssmith.wicomico.wikispaces.net/Rock+Cycle,
http://creativecommons.org/licenses/by/3.0/legalcode,CC by 3.0, 69.

Patrick Nurre has been a rock hound since childhood and has an extensive rock, mineral and fossil collection, having collected from all over the United States. In 2005, he started Northwest Treasures, which is devoted to designing geology kits for schools. He conducts numerous field trips each year in Washington State to such places as the Olympic Peninsula, Mt. Rainier, Mt. St. Helens, the Channeled Scablands, Mt. Baker and Whidbey Island. In addition, he also gives field trips to the volcano loop of Oregon and California, Mt. Hood volcanic area (Oregon), the eastern badlands of Montana and Yellowstone National Park. In 2012, he opened the Geology Learning Center in Mountlake Terrace, WA. He is a popular speaker at homeschool conventions, schools, and churches. Patrick currently co-pastors a church in the Seattle, Washington area.

If you would like to contact Patrick about speaking or field trips: northwestexpedition@msn.com
For a list of speaking topics: NorthwestRockAndFossil.com

Made in the USA
Columbia, SC
17 June 2017